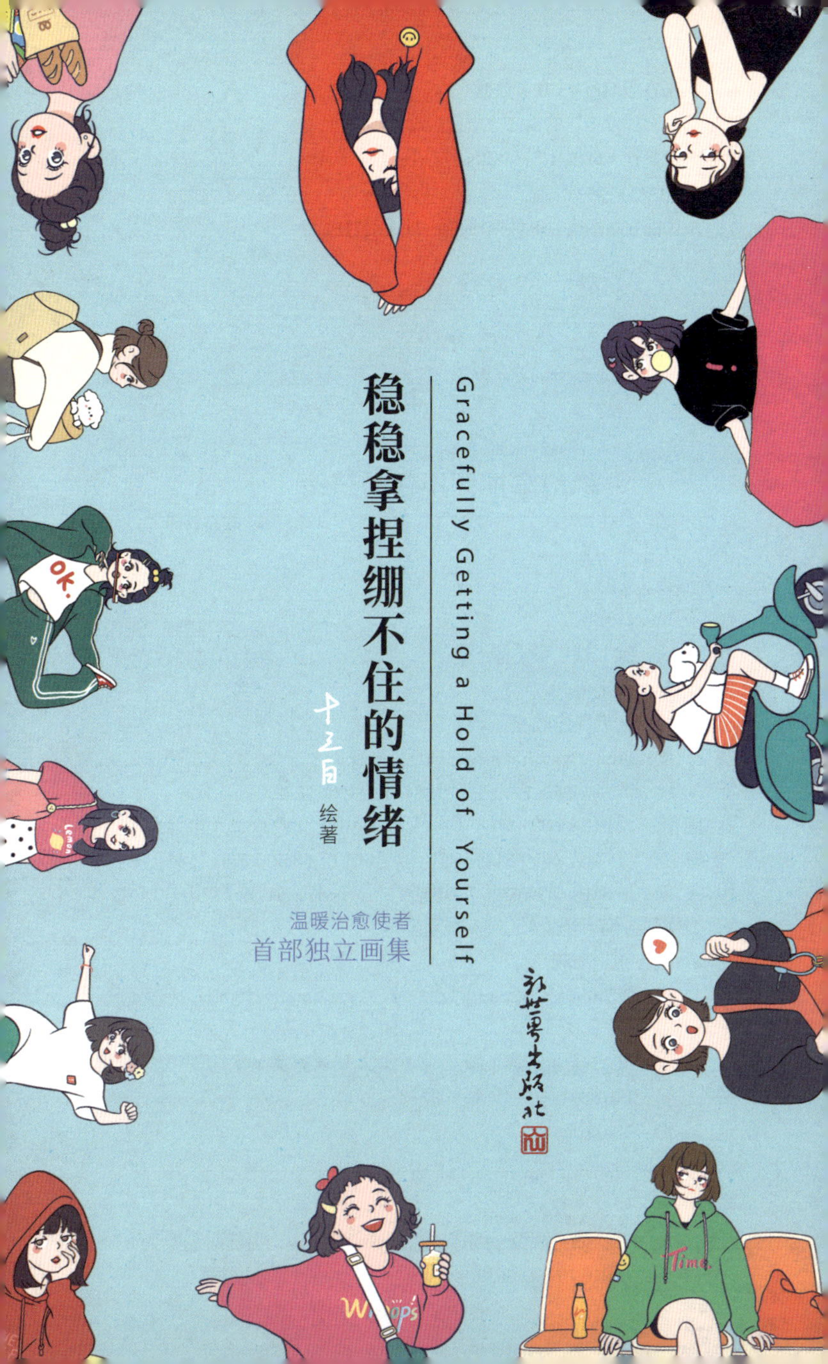

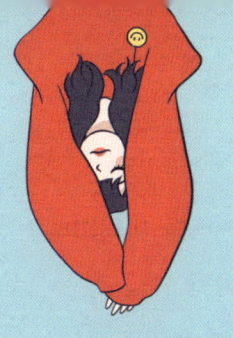

# 稳稳拿捏绷不住的情绪

Gracefully Getting a Hold of Yourself

十二月 绘著

温暖治愈使者
首部独立画集

**图书在版编目（CIP）数据**

稳稳拿捏绷不住的情绪 / 十三白绘著. -- 北京：新世界出版社, 2025.4. -- ISBN 978-7-5104-8087-4

Ⅰ. B842.6-49

中国国家版本馆 CIP 数据核字第 20250Q71F1 号

---

## 稳稳拿捏绷不住的情绪

作　　者：十三白
策划编辑：孙晓敏
责任编辑：孙晓敏
责任校对：宣　慧　张杰楠
责任印制：王宝根
出　　版：新世界出版社
网　　址：http://www.nwp.com.cn
社　　址：北京西城区百万庄大街 24 号（100037）
发 行 部：（010）6899 5968（电话）　（010）6899 0635（电话）
总 编 室：（010）6899 5424（电话）　（010）6832 6679（传真）
版 权 部：+8610 6899 6306（电话）　nwpcd@sina.com（电邮）
印　　刷：小森印刷（北京）有限公司
经　　销：新华书店
开　　本：787mm×1092mm　1/32　尺寸：120mm×180mm
字　　数：60 千字　　　　　　　　印张：12
版　　次：2025 年 4 月第 1 版　　2025 年 4 月第 1 次印刷
书　　号：ISBN 978-7-5104-8087-4
定　　价：78.00 元

---

版权所有，侵权必究
凡购本社图书，如有缺页、倒页、脱页等印装错误，可随时退换。
客服电话：（010）6899 8638

# 自 序

咖啡自由日！
A day of coffee freedom.

你好，我是插画师十三白。在数字绘画的璀璨星河中，我和你一样，是一颗正在努力发光的微尘。

我们是审美同频的旅人，才会有幸相遇。

还记得儿时，我第一次握住画笔，仿佛握住了通往新世界的钥匙。从在草稿纸上信手涂鸦，到在数位板上精心创作，绘画早已成为我生活不可或缺的一部分。我也未曾想过，这将成为我一生热衷的事情。

创作时，我常沉浸在自己的小世界里，甚至有时为捕捉瞬间灵感废寝忘食。这些画不仅是作品，更是我情感的寄托。在忙碌的尘世，画画是心灵的避风港，让我能沉浸于纯粹美好。

感谢一直支持我的人，是你们的鼓励让我坚持至今。希望翻开这本画集的你，能走进我的色彩世界，感受我对绘画的炽热，在画中找到触动心灵的瞬间。也愿它能鼓励同样热爱绘画的你，勇敢拿起命运这支画笔，不要拘束，大胆尝试，尽情挥洒属于自己的线条和色彩。

2025 年 1 月 天津

我想,
这就是我画画的意义。
素未谋面,
但可以清晰地触碰到远方的你。

This, I think, is the meaning of my art:
though we have never met,
I can reach across the distance and touch you
with perfect clarity.

# 防崩"结界"好搭子
## 治愈"名画"来了

纷扰的世界，适合独处。

In a world full of noise, solitude is the perfect refuge.

明确的爱，直接的厌恶，
真诚的喜欢。

Love with clarity, scorn with truth,
and cherish with sincerity.

坦荡地站在太阳下，
大声无愧地称赞自己。

Beneath the sun, stand unashamed,
and sing your praises freely.

# 书上说人要长大三次。

A book once said that we grow up three times in life.

# 第一次是在发现自己不是世界中心的时候。

The first is when we realize we are not the center of the world.

# 第二次是在发现即使再怎么努力，终究还是有些事令人无能为力的时候。

The second is when we understand that no matter how hard we try, some things will forever be beyond our grasp.

# 第三次是在明知道有些事可能会无能为力，但还是会尽力争取的时候。

The third is when, despite knowing the limits of our power, we still choose to fight for what matters.

"日日是好日"留言板

我发现一个人在放弃给别人
留下好印象的负担之后,
原来心里会如此踏实。

I discovered that when you relinquish the burden of trying to leave a good impression on others, your soul finds a profound sense of peace.

快乐使肉身美丽，
痛苦让灵魂升华。

Joy beautifies the flesh;
pain refines the soul.

我除了年龄以外,
一点也不像个大人啊!

Other than my age,
there's little about me that resembles an adult!

人要是只剩下了一个选择，
路倒是更好走了，
越走胆儿越大。

When all that's left is a single choice, the road becomes simpler—and your courage grows with every step.

当你穿过暴风雨,
你就不是原来的那个你了。

After walking through the storm,
you are no longer the person you once were.

你若要彩虹，
就得先接纳风雨。

If you yearn for the rainbow,
you must first embrace the rain.

减肥日记：该丢的脸丢过了，人也不好意思不成长了。
Diet journal: I've lost all the face I could lose—there's no excuse not to grow.

我别无良策，唯耐心以赴。

I have no clever solutions, only patience to see me through.

第一是健康，事业次之，情啊爱啊，好自为之。

Health comes first, career second;
love and passion, let them fall where they may.

很爱我时常普通，
偶尔糟糕的小生活。

I love my simple, often messy,
and occasionally chaotic ordinary life.

稳稳"晴雨表"

| 开心 | 平和 | 喜欢 | 期待 | 生气 | 惊讶 | 悲伤 | 未知 |
| --- | --- | --- | --- | --- | --- | --- | --- |
|  |  |  |  |  |  |  |  |

我度我自己。

I am the one who saves myself.

手持烟火以谋生，心怀执着以谋爱。

With hands full of humble firelight to make a living, and a heart steadfastly burning to seek love.

常与同好争高下，
不共傻瓜论短长。

Strive with your equals,
but never waste words with fools.

## 以一个敞开的灵魂去关注一切，
## 没有狂躁，没有期盼，没有论断，没有成见。

Approach the world with an open soul—free from frenzy, expectation, judgment, prejudice.

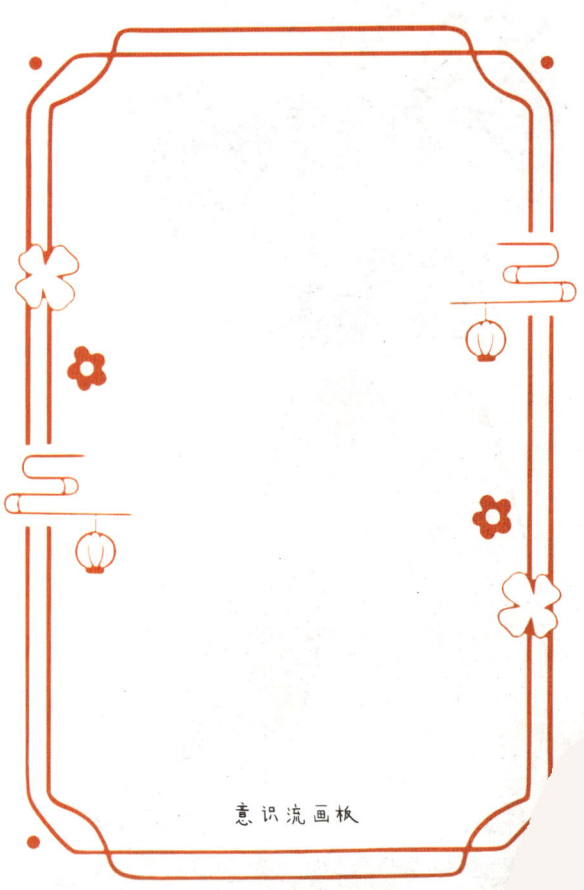

意识流画板

不要用别人的错误惩罚自己。
来，笑一个！

Don't punish yourself for the mistakes of others.
Come now—smile!

我曾经以为日子是过不完的,
未来是完全不一样的。

I once believed time stretched endlessly,
and the future would be entirely different.

现在,我就待在我自己的未来,
我没有发现自己有什么真正的变化,
我的梦想还像小时候一样遥远,
唯一不同的是我已经不打算实现它了。

Now I stand in the future I dreamed of,
but nothing within me has truly changed.
My dreams remain as distant as when I was a child,
and the only difference is that I no longer intend to achieve them.

过去的事情是因为熟透了才从你身上掉下去。
曾以为走不出的日子，现在都回不去了。

**Things from the past fall away only because they've ripened.
The days I thought I could never escape have become days
I can never return to.**

人生最忌满，不能太完美，
半贫半富半糊涂。

The gravest flaw of life is perfection.
Let us remain half-needy, half-wealthy, and half-serene.

"日日是好日"留言板

焦虑在敲门,勇气打开门,
门外什么都没有。

Anxiety knocks at the door;
courage opens it, only to find nothing
waiting outside.

清喜于时光，安稳于日常。

Find quiet joy in time, and steadiness in daily life.

万物都有节奏,
小闲即欢,小清即静。

Everything has its own rhythm—
a little leisure brings joy, a little quiet brings peace.

不如就承认一下，我们没有那样坚强，也不想那样刀枪不入，我们只是想被温暖地抱一下。

Perhaps it's time to admit: we are not as strong as we pretend to be, nor do we wish to be invincible. What we truly long for is a warm embrace.

躲了一辈子的雨,雨会不会难过?

I have spent my life running from the rain— has the rain ever felt sorrow?

重要的人越来越少,
剩下的人也越来越重要。

**The truly important people grow fewer with time,
and those who remain become all the more significant.**

药在时间里。
The cure lies in time itself.

日进斗金，细水长流。

Fortune flows in daily increments,
steady and enduring like a gentle stream.

你不能等到生活不再艰难，
才决定开始快乐。

You cannot wait until life becomes less difficult to decide to start being happy.

此刻小情绪 | 开心 | 平和 | 喜欢 | 期待 | 生气 | 惊讶 | 悲伤 | 未知

斯人若彩虹，
遇上方知有。

Rare is the person who is like a rainbow;
only after meeting them do you realize their wonder.

39

穿舒服的衣服，和松弛的人相处。

Wear what makes you comfortable, and get along with relaxed people.

中年少女，有小肚腩，性格带刺，那又怎样？我们是野蛮生长，永怀希望的仙人掌。

Middle-aged women with stubborn bellies and thorny personalities—so what? We grow wild like cacti, tough yet eternally hopeful.

41

天上的星星
笑地上的人。

The stars in the sky gaze down with quiet amusement at the people on the earth.

晒自己。

Show yourself to the world—fearlessly and unapologetically.

莫待

花谢空折枝

Actively love

前途与玫瑰，来日与方长。

Ahead lie roses and the promise of future.

一键消除烦恼

断舍离。

Declutter. Simplify. Let go.

日日落俗，日日开心，不过是一场生活。

Day after day, we follow the routine and find joy; such is life.

如果早认清我在别人心中没那么重要，
我会快乐很多。

If I had realized earlier how little I truly mattered to others,
I would have found much more happiness.

这个世界看似纷繁复杂，
其实本质却还是我一个人的世界。
真心做个快乐的自己吧！

This world seems complex, yet at its core, it's my world alone.
Be truly myself and live happily.

矛盾的是，生活只能向从前理解，
但它又必须向未来发展。
人无法同时拥有青春和对青春的感受。

**Life is a paradox: understood in hindsight, yet lived forward. Youth and its awareness cannot coexist.**

为了记住你的笑容,
我拼命地按下心中的快门。

In a desperate attempt to preserve your smile,
I pressed the shutter of my heart with all my strength.

50

我常常反问,
为什么善良的人生总是一波三折,
而自私的人生却是一顺再顺?
或许我这样想本就不对。

I often ask myself:
why do kind-hearted lives stumble through twists and turns,
while selfish lives glide along effortlessly?
Perhaps my thinking itself is flawed.

我要相信,
人生就是一场修行,要度劫,
无论正在经历着什么,
结局一定都是美的。

I must believe that life is a journey of spiritual cultivation,
a trial to be endured.
No matter what we are experiencing now,
the ending will surely be good.

"日日是好日" 留言板

长久以来，我总觉得新生活就要开始了，
只是总有障碍挡道，有麻烦要克服，
有事情还未完成。

For so long, I believed my new life was just about to begin.
Yet there were always obstacles to overcome,
troubles to resolve, and tasks to finish.

我总认为只有把障碍清除了，把麻烦解决了，把事情做完了，新生活才能真正开始。

I thought that only by clearing these hurdles could I truly start a new life.

这种想法去去来来，一直在重复，
但新生活却没见轰轰烈烈地开始。
终于，我突然明白，
原来这些障碍本身就是生活。

This thought repeated itself endlessly,
yet the life I envisioned never arrived.
And then I realized these obstacles were life itself.

什么是生活？就是逼得你无路可走时，
给你一点点希望，让你继续走下去。

What is life? It is what offers you a sliver of hope when all paths seem closed, urging you to press on.

53

这是一个流行离开的世界，
但是我们都不擅长告别。

This is a world where departure has become a part of everyday life, yet we are all clumsy in saying goodbye.

成年人的告别仪式很简单，你没有回我消息，我也便默契地没有再发，就这样安静地消失在彼此的生活里，好像从没认识过一样。

The farewell of adults is simple—if you don't reply, I won't insist. And so, we quietly vanish from each other's lives, as if we had never met at all.

别人是会离开的，但你自己一直都在。照顾好自己，好好爱自己。

Others may come and go, but you will always be there for yourself.

In the end, it is you who must care for and cherish your own soul, for self-love is the most enduring companion.

人没必要一直都是大度的，
我们也可以很自私，
我们不是为了让大家称赞才活着的。

**Generosity isn't an obligation.
Prioritizing yourself is completely acceptable—after all,
we weren't born to seek others' validation.**

小时候不懂，嘲笑大人听戏的样子真老土。

As a child, I laughed at how adults seemed so old-fashioned when they listened to opera.

直到有一天，开着车听到了广播里的一段戏腔，回到家到处搜索到底是哪段的时候……才发现，初闻不知曲中意，再听已是曲中人。

Until one day, while driving, I heard a fragment of opera on the radio. When I got home, I searched everywhere to find out what it was. That's when I understood: "The meaning of the melody escapes me at first, but on the second hearing, I find myself in its story."

奇怪，小时候明明觉得戏曲很难欣赏，甚至听了会犯困，为什么现在就听进去了呢？原来……这就叫作"老去"吧。

It's strange—what once seemed incomprehensible now resonates deeply. Is this what it means to grow older?

没有任何一种逃避能得到赞赏。
喜欢就去追寻，饿了就吃饭，
管它失败或是发胖。

**Running from choices—even small ones—won't get you claps!
Chase what you desire, eat when you are hungry,
and never mind failure or gaining weight.**

要么读书，
要么赚钱。

Either read books
or make money.
There is no third option.

万事都是自由诗。

Everything in life is free verse.

别忘了抽时间犒劳自己。
Don't forget to take time to pamper yourself.

人生，四十岁才刚开始。
除了生死，没什么大事。

Life truly begins at forty.
Beyond life and death,
there is nothing truly worth fretting over.

永远不要去美化你没有选择的那条路。
Never romanticize the path you didn't choose.

Extroversion is a survival skill; solitude is the reward of the self.

外向是生活所需，孤独是自我享受。

梦想养不起你的时候,
你得养着梦想啊!

When your dreams can no longer sustain you,
it's your turn to sustain them.

小时候，幻想自己会是一个仗剑走天涯的女侠。

As a child, I imagined myself as a swordswoman roaming the land.

长大后，手里没有剑，只有市井中的大葱。

As an adult, my hands hold no sword—only a leek from the market.

等你发现时间是贼，
它早已偷光你的选择。

By the time you realize time is a thief,
it has already stolen all your choices.

这个世界,没有一种痛是单独为你准备的。

No pain in this world is reserved solely for you.

不要认为你是孤独的疼痛者,也不要认为,自己经历着最苦的痛。

Don't think of yourself as the lonely bearer of suffering, nor believe your pain to be the deepest.

尘世的屋檐下,有多少人,就有多少事,就有多少痛。

Beneath the eaves of this earthly world, there are as many sorrows as there are people.

在芸芸众生的痛苦里,你会发现,自己的这点痛,可能并不算什么。

And in this vast ocean of human grief, you may come to see that your own pain, in comparison, is but a fleeting wave.

虽然不知道被困在哪里，
但我想冲出去。

**I don't know exactly where I'm trapped,
but I know I want to break free.**

纯爱满足不了我，纯金可以。

Pure love can no longer satisfy me—but pure gold might.

我不躲不藏，等生活爱我。

I'm neither hiding, nor running. I'll wait for life to love me.

## 有些瞬间会支撑我。

**Some moments will always hold me up.**

此刻小情绪 | 开心 平和 喜欢 期待 生气 惊讶 悲伤 未知

"静"字中尚且还有一个"争"字,它是要在世界的绚烂和复杂中,奋力夺来的。

The Chinese character for "stillness"—静—hides within it the word "strive"—争. It is something wrested from the chaos and brilliance of the world.

永远不会有人完全理解你，
认识到这点就是自由。

No one will ever fully understand you.
Realizing this is the first step to true freedom.

冰冻三尺非一日之寒，
小腹三层非一日之馋。

A river does not freeze in a single day,
and a soft belly is not the result of a single indulgence.

生活要埋了我,却不料我是颗种子。

Life intended to bury me—little did it know, I am a seed.

人性有恶。
有时也正是这"恶",能保护自己。

**Humanity harbors darkness.**
**Yet sometimes, it is this very darkness that keeps us safe.**

# 得闲幸事，八九不离食。

To idle in leisure and delight—
nothing strays far from the joy of food.

以后能用酒解决的事,就不麻烦眼泪了。

From now on, let wine resolve what it can, and let tears rest.

你无法安慰一个心如明镜的人，
因为她太懂了。

You cannot console someone whose heart is as clear as a mirror—for they understand too much already.

"瘦"字是病字头，
所以"胖"才是健康哒！

The Chinese character for "thin" begins with the radical for illness, so "fat" must be the symbol of health.

我不是胖，
只是可爱到膨胀。

**I'm not fat—
I'm just so adorable that I've expanded.**

情绪还是在自己的手里安心。

Keep your emotions in your own hands—
it's the safest place for them.

我这白开水一般的生活,需要加一点甜。
My plain, water-like life should add a little sweetness.

迷路，也是走路的一部分。

Losing your way is just another step in finding it.

LOSE ONE'S WAY

稳稳"晴雨表"

| 开心 | 平和 | 喜欢 | 期待 | 生气 | 惊讶 | 悲伤 | 未知 |

无事发生，就是最好的一天。

A day when nothing happens is the wonderful day.

情绪提醒你，
是为了让你去解决产生的问题，
而不是令你纠结于情绪本身。

Emotions are messengers—they remind you to face what's broken, not to drown in the weight of feeling itself.

人一辈子，
走走瞧瞧，
吃吃喝喝，
不生病，
就是福气。
如果能遇到自己爱的
也爱自己的人，
再发点小财，
就是天大的福气。

Life is simple:
wander,
explore, eat, and drink.
Staying healthy is a blessing.
If you could find someone you love who loves you back,
and stumble upon a bit of fortune, that's the kind of
happiness you could call divine.

三餐四季,
须臾一载。
不管生活怎么辜负,
妈妈都不后悔生下你,
因为你是妈妈崩溃后唯一看得见的光,
妈妈爱你。

Three meals, four seasons,
and years that slip by in the blink of an eye.
No matter how unkind life has been,
a mother will never regret having you.
You are the light that remains after her darkest days.
You are the love she cannot live without.

突然发现厚脸皮根本没用，
风吹过来冻得半死。

**Thick-skinned is useless at all—
the wind will still cut through to the bone.**

# 我要给你一个金钟罩，
## 一个铁布衫！

**I want to wrap you in golden armor, shield you with steel, and protect you from the world's cruelties.**

"日日是好日"留言板

365个祝福旺自己一整年

## 疲惫且努力地奔向周末。

Tired yet relentless, I push forward, rushing toward the weekend like a desperate runner reaching the finish line.

黎明之前最黑暗，
放弃就是大笨蛋。

The darkest hour comes just before dawn—
only a fool would surrender when light is so near.

人心各有所愿，是没有道理可讲的。
如果快乐太难，那我祝你平安。

The heart wants what it wants—there's no reasoning with it.
If happiness feels out of reach, then I wish you peace,
pure and simple.

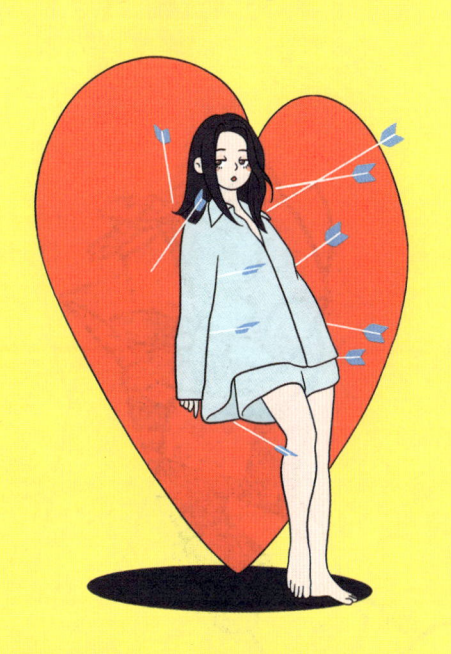

花朵从不慌张。

Flowers never hurry—they bloom when they're ready.

幸福是无数个瞬间的总和。

Happiness is made up of countless fleeting moments.

热爱，就是在混沌中，
永无止尽地挥棒，挥到死。

True passion is swinging the bat endlessly into the chaos,
till the end of life.

风住风往，
接受生活的
每一次
不如意。

The wind will rise and fall.
Learn to embrace
the disappointments life
sends your way.

本想一口一口吃掉忧愁，
没想一口一口吃成肉球。

I thought I could eat my sadness away, bite by bite. Instead, I ate my way into a rounder, softer version of myself.

即使说了那么多丧气话，
也一直在拼尽全力地生活呀！

Even after all my complaints and hopeless words,
I've still been striving for life.

看完喜欢的小说或剧,
就像送别了一帮自己的朋友。
我剧荒了。

Finishing a beloved book or series feels
like saying goodbye to a group of close friends.
Now I'm left stranded in the silence they've left behind.

### 你知道吗，
### 疼痛是这个世界对你的挽留。

Pain, strange as it sounds,
is the world's way of holding onto you, asking you to stay.

**稳稳"晴雨表"**

| 开心 | 平和 | 喜欢 | 期待 | 生气 | 惊讶 | 悲伤 | 未知 |
|---|---|---|---|---|---|---|---|
|  |  |  |  |  |  |  |  |

## 只有走起来，才知道方向。

Only by moving forward can you begin to see the path.

人一出生就会哭，笑是后来才学会的。
所以忧伤是一种低级的本能，
而快乐是一种更高级的能力。

We are born with the ability to cry,
but laughter comes later—it has to be learned.
Sorrow is instinct;
joy is a hard-earned skill.

墨镜一戴，谁也不爱。

Put on your sunglasses, and do not care about anyone.

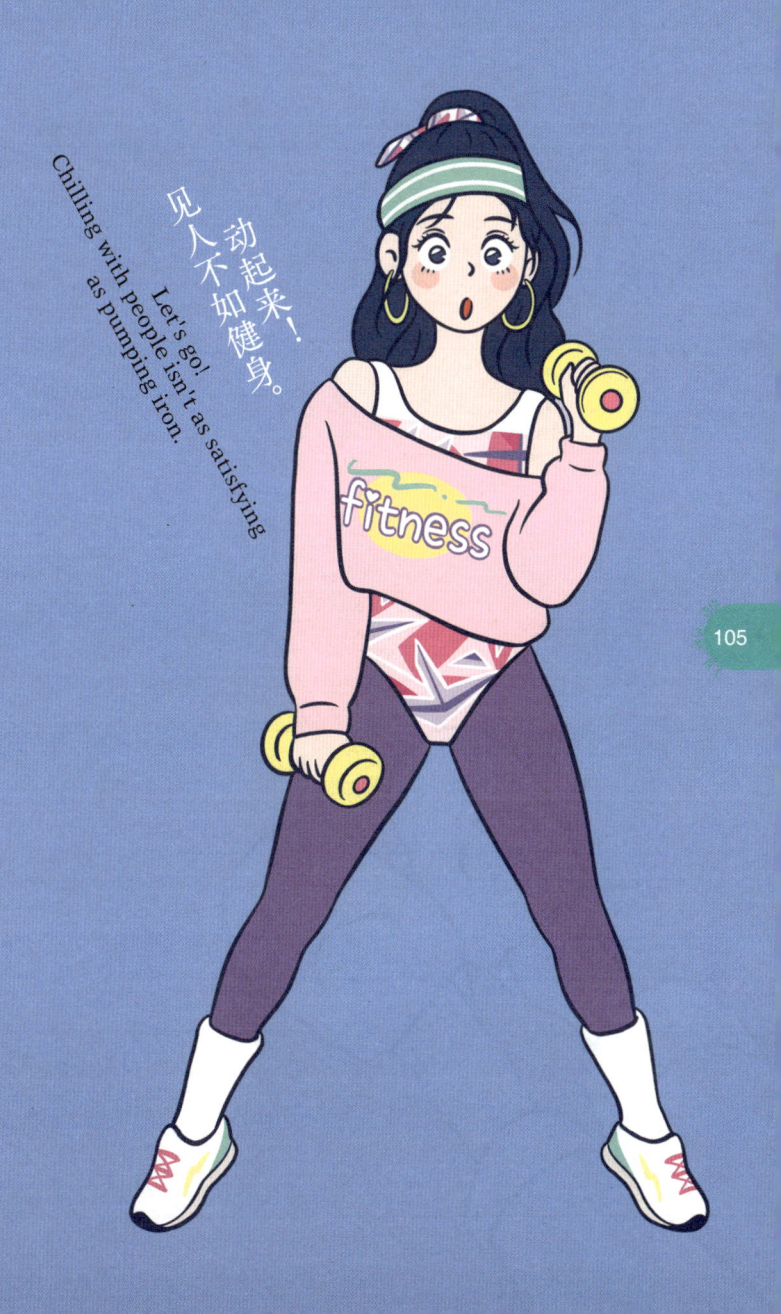

动起来！见人不如健身。
Let's go! Chilling with people isn't as satisfying as pumping iron.

105

被市井煮沸，但我依旧平静。

Even when boiled in the hustle and bustle of the marketplace,
I remain calm and still.

慢慢理解世界，慢慢更新自己。

Gradually understand the world, and renew yourself.

看穿所有的虚伪，但我仍然真诚。

人间清醒。
生活，就是从一堆碎玻璃碴子里找糖。

Clarity is bittersweet:
life is searching for sweetness amidst shards of broken glass.

有被讨厌的勇气，
人生就自由了。

Having the courage to be
disliked leads freedom in life.

无意中看到一个网友的情绪树洞，
她说，她初中的好朋友，
到现在还在和当初的好朋友一起玩，
一起旅行。

I came across someone's online confession.
She wrote about that one of her close friend in middle school still hangs out and travels with their middle school friends.

突然就反思自己，
为什么每一个阶段身边的人都留不住？
……

And I wondered:
why can't I seem to keep friends from each chapter of my life?
……

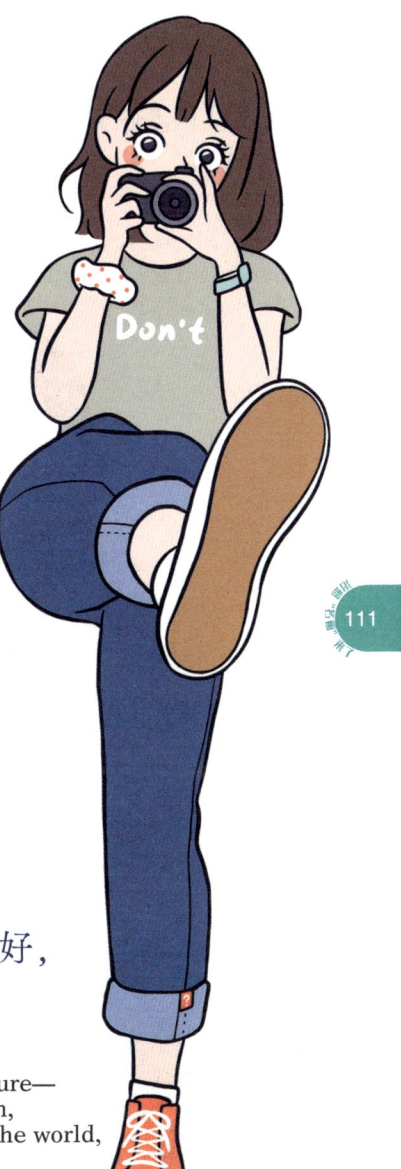

希望
我能成为你的小众喜好,
藏着欣喜不已,
炫耀时格外骄傲。

I want to be your hidden treasure—
a quiet joy you secretly cherish,
and when you show me off to the world,
your pride glows.

小时候觉得，独来独往的人很古怪。
长大后才发现，他们活在更大的格局里。
学会与自己独处，是最难的事。

As a child,
I thought solitary people were strange.
As an adult,
I see that they're the ones who live with the greatest clarity,
unbound by the expectations of others.
Learning to live with yourself needs the greatest courage.

# 内向而不呆滞，寂静而有力量。

**Introverted, but not sluggish; Silent, but powerful.**

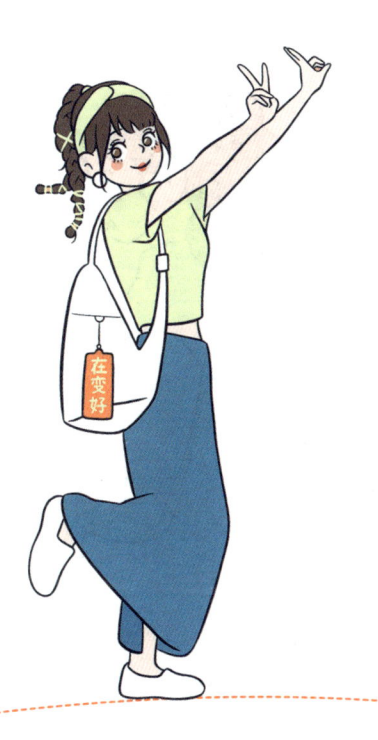

"日日是好日"留言板

没有期待的日子，反而顺顺利利。

**Days without expectations often go smoothly.**

## 真希望所有的事，都像长胖一样简单。

I wish everything in life could be as effortless as gaining weight.

如果累 = 瘦，就好了。

If exhaustion equaled weight loss,
life would be so much easier.

我心向光，
何惧悲伤。

My heart turns toward the light,
unafraid of sorrow.

发财，被爱，好运常在。

Wealth, love, and endless good fortune—that's all I ask for.

悲喜自度，他人难悟易误。

Joys and sorrows are one's own to navigate,
and others may find them hard to comprehend.

此刻小情绪 | 开心 | 平和 | 喜欢 | 期待 | 生气 | 惊讶 | 悲伤 | 未知

问题不大。

It's no big deal.

人生的旅程就是这样，
用大把时间迷茫，
在几个瞬间成长。

Life's journey is like this:
we spend so much time in confusion,
and then grow in fleeting and precious moments.

世上的事情都经不起推敲,
一推敲,哪一件都藏着委屈。
我的所有快乐,
总是源于我的健忘和喜欢的一切。

Nothing in this world can withstand too much scrutiny.
Every matter hides its own grievances.
My happiness, fragile as it is,
always comes from my ability to forget and my love for the little things.

摸摸小油头，万事全不愁。

Pat my greasy little head, and all my troubles fade away.

停止向任何人解释你自己的生活。
Stop unveiling the story of your existence to others.

世界好吵，想躲起来…

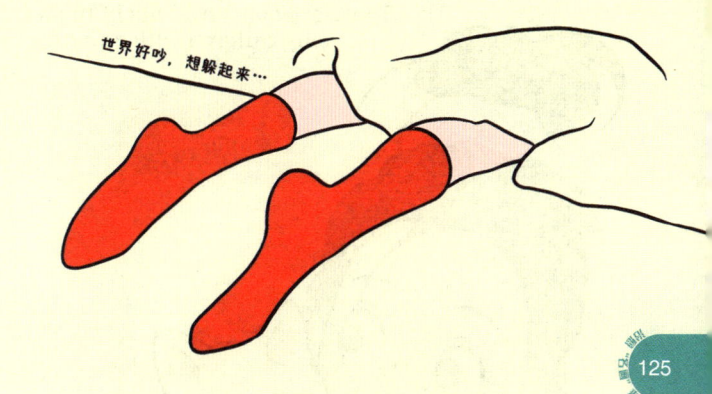

请允许我枯萎几天。

Please, allow me a few days to wither.

"日日是好日"留言板

这个世界根本没有大人，
只有长皱了的小孩。

There are no true adults in this world—
only children with wrinkles.

没意义的人或事就该丢掉，
拎着垃圾的手怎么腾得出来接礼物？

**Meaningless people and things should be discarded.
How can your hands hold gifts if they're full of trash?**

我不需要你来告诉我,我应该是什么样子。
你不喜欢我,那不是我的错。

I don't need you to tell me how I should be.
If you don't like me, that's not my fault.

每个人都在等待些什么。
等一个人回头，等一个人出现，等自己释怀。
等春暖花开，等灯火通明。
如果能等到自己想要的，就没有浪费时间。

Everyone is waiting for something—
a person to turn back, a new face to arrive,
one's own heart to find peace,
a spring to bloom or a street to be illuminated.
If what you're waiting for finally comes, then no time was wasted.

绝不试图从感情废墟中
刨回任何断壁残垣。

Never seek to unearth the crumbled remnants from the wreckage of lost emotions.

按你所想的去生活。
否则,迟早会按你生活的去想。

Live the way you want to, or someday you'll find yourself trapped, trying to want the life you settled for.

此刻小情绪 | 开心　平和　喜欢　期待　生气　惊讶　悲伤　未知

答案，在硬币的第三面。

The answer lies on the third side of the coin.

愿你健康，开心，总有好运气。

May you be healthy, happy, and lucky in all things.

接受一切遗憾。
Accept all the regrets life has left you.

恰到好处描述今天。

Find the perfect words to describe today.

135

不应该为了一个瞬间，放弃人生无数个瞬间。

One should not sacrifice countless moments of life for a single instant.

意识流画板

我感受到的不舒服就是不舒服。
我感受到的不友好就是不友好。

Discomfort, when I feel it, exists genuinely.
Unfriendliness, when it stings, stands as it is.

我要直观地承认自己的感受,
用不着过度剖析他人,反思自己。

I choose to acknowledge my feelings directly,
without excessive analysis of others or self-reflection.

她昨天骂了我一声"笨蛋",我一笑而过。

Yesterday, she called me "stupid", and I let it slide with a smile.

今天我也说了她一句"笨蛋",不知道为什么她就生气了。

Today, I called her "stupid", and for some reason, she got angry.

我只是用了她对待我的方式对待她呀……

I only treated her the way she treated me—why does she feel so unfair?

任何瞬间的心动都不容易，
不要怠慢了它。

A fleeting moment of heart-stirring emotion is rare—
don't take it for granted.

谁不是玻璃心啊，只不过是在心上裹了厚厚的消音布，让稀里哗啦碎一胸腔的声音不被别人听见，自己也假装听不见而已。

Who doesn't have a fragile heart?
We just wrap our hearts in layers of soundproof fabric,
so when they shatter into tiny shards, no one hears it.
Not even ourselves.

我好像来错星球了,请带我回去吧。
如果不行,我就和这烂糟的世界绝交!

I think I've landed on the wrong planet. Please, take me home. And if that's not possible, then I'll sever ties with this messy, chaotic world!

春天一到，一切忧虑都随即消散，即使是夜晚。

Spring has arrived, and with it, all my worries have melted away—
even the ones that linger at night.

拒绝 emo!

No more emo vibes—I refuse!

阳光的味道，好像时光机。

The scent of sunshine has the magic of a time machine, taking me back to reminisce about the past.

今日阳光甚好，快出来晒晒吧！♥

意识流画板

糟糕，坠入爱河！
Oh no—I've fallen in love!

何必把怀念弄得比经过还长。

**Why stretch nostalgia longer than the memory itself?**

天气暖和起来啦，脱掉厚重的衣服，突然有点失去安全感。

The weather is warming, and I've shed my heavy winter layers. But suddenly, I feel exposed—like I've lost the comforting security.

好吧，明天我会轻装上阵，适应变化，努力勇敢。

It's okay. Tomorrow, I'll face the world lighter, ready to adapt and brave whatever comes my way.

我始终相信在这个世界上，一定有另一个自己，做着我不敢做的事情，过着我想要的生活。

I've always believed that somewhere in this world, there's another me, doing the things I dare not do, living the life I've always dreamed of.

少与人纠缠，多看大自然。

Engage less in human entanglements, and immerse yourself more in nature.

致敬这必死无疑的一生。

Salute to this life that is doomed to die.

你是不是也觉得，为什么每个人看起来
都过得那么好的样子，
只有你，糟糕得不像话……
（错觉！完全是错觉！）

Do you ever feel like everyone else seems to have lived a good
life while yours is a complete disaster?
( It's an illusion! Nothing but an illusion! )

我不怕掉眼泪，但要值得。

I don't mind shedding tears, as long as it's worthwhile.

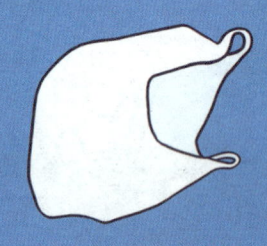

沉默的
胆小的
普通的

世人的偏见是座山，
但我偏要一脚踢碎它。

The world's prejudices rise like a mountain,
yet I shall kick them away with one swift move.

即使我见过很多复杂和阴暗，
但我依然不屑成为那样的人。

Even though I have witnessed much complexity and darkness, I choose to remain untainted.

# 吃瓜，吃瓜！

Just here for the drama—pass me the storylines!

或许很多时候，记忆是最好的见面方式。

Perhaps, at times, memory is the best means of reunion.

"日日是好日" 留言板

爱是一回事，生活是一回事，
心动是一回事，岁月是另一回事。

Loves doesn't always undergo the tests of life.
A momentary infatuation doesn't necessarily last through the years.

空气清新得像一场大病初愈。
放宽心情，什么都变美了。

Air whispers freshness like a soul rising from the grave of sickness.
In releasing tension, the world reveals its hidden radiance.

去喜欢一个让你有动力的人吧，
每天起来都觉得晴空万里；
而不是喜欢一个让你有伤口的人，
每天睡去都觉得疼痛无比。

Choose someone who fuels your drive, making every dawn feel like clear skies, rather than someone who leaves scars, turning every dusk into agony.

我虽然并没有忘记你，
但也没有精力再去想了。

Though I haven't forgotten you,
I no longer have the energy to think about you.

好羡慕地板啊，
有那么多头发。

I envy the floor;
it has so much hair.

这里的每一个词语,
都是一生中最奢侈的愿望。

Every word here is the most extravagant wish in a lifetime

平安喜乐!
大吉大利～
發財
以好运
健康
萬事顺意↑
平♡安
加油!

刚刚看到一句话说:"人一旦安于现状很久后,便会少了出逃世界的勇气。"

I have read a line: "Once you've stayed in your comfort zone for too long, you lose the courage to escape it."

可我觉得,安于现状也需要勇气,不是吗?

But it also takes courage to be content with the status quo.

出逃世界就一定是标准答案吗?

Is escaping the world always be the standard answer?

比起偶然的幸福感，
充足的睡眠似乎有着更持久的治愈力。

**Sleep is the quiet cure that lasts far longer than fleeting bursts of happiness.**

做自己的云朵，柔软且温暖。

Embrace yourself like a cloud, tender and warm.

纵你阅人何其多，
再无一人恰似我。

You may know countless people,
but none will ever be like me.

我把梦想装进了纸船，它漂得很远很远……
直到有一天我彻底把它遗忘。
世事无常，我不是故意的，希望你原谅我。

I packed my dreams into a paper boat and sent it floating far, far away, until I forgot it one day.
The unpredictability of life distracted me from my aspirations. It wasn't on purpose; I seek forgiveness from my dreams.

亲人离世，就是你在学校，他在赶集；
你在家里吃饭，他在地里干活；
你去地里找他，他又恰好回到了家；
他永远都在，
只是今后每次都会擦肩而过……

Losing a loved one feels like this: you're at school, and they're at the market.
You're eating at home, and they're working in the fields.
You go looking for them in the fields, only to find they've returned home. They'll always be there, but from now on, you'll only pass by each other, never truly meeting again.

有一种女人，内心强大得像开玩笑似的。
售货员说这件衣服500元，
她可以直接说出：“50元卖不卖？”

There's a kind of woman whose inner strength is ridiculously strong. When told a dress costs 500 yuan, she simply replies, "Will you sell it for 50?"

170

亲爱的中年少女，你过得还好吗?
只要皱纹不长在心里，我们永远风华正茂。

**Dear middle-aged woman, how are you?
As long as the wrinkles don't reach your heart,
you'll always remain radiant.**

171

意识流画板

今天，我对着镜子发呆，眼前陌生的女子，我似乎很久没认真端详了。

几滴眼泪不听使唤地滴了下来。
平复心情之后，我努力尝试原谅并接受自己。
这几滴莫名其妙的眼泪就当是礼物吧！

致敬新长出来的两条细纹。
致敬失联已久的胶原蛋白。
致敬黑发丛中的白色翘楚。
致敬十年都没再穿过裙子的那个——中年少女。

Today, I stared at the mirror, and the woman who looked back at me felt like a stranger.
I realized it's been so long since I truly looked at her.

A few tears escaped, unbidden.
After calming down,
I strive to forgive myself for overlooking my personal changes and accept the changes in my appearance as I age.
These tears? I'll take them as a gift.

A tribute to the fine lines that have appeared.
A tribute to the long-lost collagen.
A tribute to the silver strands hiding among the black.
And a tribute to myself who hasn't worn a dress in ten years.

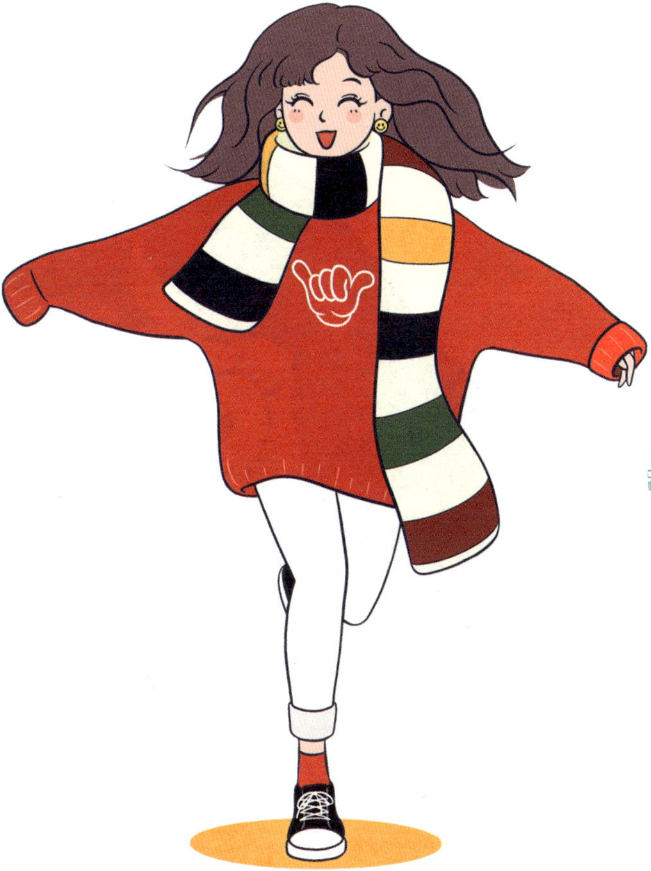

大部分人和你已经见完了此生的最后一面。

Most people have already seen you for the last time in this life.

今天瑜伽不狠，明天地位不稳。
No pain in yoga today, no gain in status tomorrow.

此刻小情绪

| 开心 | 平和 | 喜欢 | 期待 | 生气 | 惊讶 | 悲伤 | 未知 |

有趣的灵魂万里挑一。

An interesting soul is one in a million.

契合灵魂同频共振。
宁缺毋滥独自美丽。

Seek the souls resonate with yours. If there are no such souls, please embrace your solitude with grace.

总想着别人的感受，
灵魂就回不到自己的身体。

**Constantly thinking about others' feelings
keeps your soul from coming back to yourself.**

不爱,是一生的遗憾。爱,又是一生的磨难。
你告诉我,到底什么是对的?
人生的意义又是什么?
我最终只得到了两个字——体验。

这世上有多少个人,就有多少种结果。
而结果,也没有任何意义。

To not love may become a regret in one's life. To love may bring a lifetime struggle.
So tell me—what's the right choice? What is the meaning of life?
In the end, I found only two words: to experience.

As many people as there are in this world, there are just as many endings. And in the end, none of it really matters.

幸与不幸，都有尽头。

Happiness and sorrow both have their limits.

删除一些难过,才能容纳一些快乐。

To make room for joy, you must first let go of sorrow.

成年人也只是过期的小朋友。
Adults are simply kids past their expiration date.

熬过低谷，不要慌不要慌，
太阳落了有月光。

Get through the low points, don't panic.
When the sun sets, the moon will shine.

睡眠不足。 Sleep-deprived.

他问我有什么打算,
下一秒的事谁知道呢……

He asked me about my plans.
Plans? Who knows what the next moment will bring?

爱与自由二八分，
失之坦然，得之淡然。

Love and freedom are split 80/20.
Accept loss with ease, and gain with calmness.

# Only ♡ Me

管住嘴,迈开腿。

Watch your mouth and move your legs.

横扫内耗，做回自己。

Eradicate self-consumption—be unapologetically yourself.

好运总是要先捉弄一番，
然后才向着坚韧不拔者微笑的。

**Good luck often plays a trick first,
before smiling upon those who persist with
unwavering determination.**

"日日是好日"留言板

你努力学习、沉迷工作的样子很迷人。

The way you immerse yourself in studying and working is quite captivating.

你精心烹饪、热爱生活的样子也很迷人。

The way you cook with care and embrace life is also charming.

速冻模式的冬天，把自己好好保存，
来年开春，我们依旧新鲜。

**Preserve ourselves well in the frozen winter.
In the coming spring, we'll renew ourselves.**

我和这个世界不太熟，
但我已经是个大人了。

I'm not familiar with this world,
though I've reached adulthood.

如果每个人都懂你，
那你得普通成什么样子。

**Being understood by all may mean you're blending in too much.**

胖胖的，也可以美美的。
拒绝容貌焦虑。

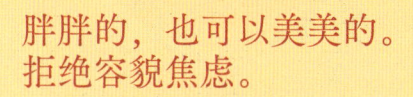

Being chubby can also be beautiful.
Let's leave appearance anxiety behind.

睡前原谅一切，白天干翻世界。

Forgive everything before sleep, and conquer the world by day.

"月亮"是隐喻，
所有的本体都是你。

The moon serves as a metaphor,
with you being the essence of it all.

无人照料的小花，也会独自盛开。

Even an untended flower will bloom on its own.

不合群,说明你没有委屈自己。
这是一件值得高兴的事情。

**Being different means you haven't betrayed yourself—that's something to celebrate.**

热烈而又自在地活着。

Live with passion and freedom.

别给我画饼了,我最近在戒碳水。

Don't give me pie-in-the-sky promises; I am on a low-carb diet.

上得厅堂，下得厨房。

Able to grace the hall and manage the kitchen.

永远不要在机场等一艘船。

Never wait for a boat at an airport.

小时候，
都觉得自己的未来闪闪发光，不是吗?

**As children,
didn't we all think our futures would sparkle?**

不可以垂头丧气哦，会显矮。

Don't sulk—it makes you look shorter.

真正的忘记，是不需要努力的。

True forgetting requires no effort.

有一天，你从浴室洗了一个澡出来，
扭开唱机听听自己喜欢的音乐，你忽而想起，
你曾经爱过一个人。
啊！原来你爱过这个人，那仿佛是很遥远的事，
你已经一点感觉也没有了。
这就是忘记。

One day, after taking a shower, you turn on the record player to listen to your favorite music, and suddenly remember that you once loved someone.
But it feels distant, as if it happened to someone else.
And you feel... nothing.
That's what it means to forget.

不走心的努力，都是在敷衍自己。

Any effort without genuine commitment is a disserrice to oneself.

此刻小情绪　　开心　平和　喜欢　期待　生气　惊讶　悲伤　未知

这个世界需要无用的东西,
什么都有意义的话,你会感到窒息。

This world needs useless things.
If everything had meaning, you might feel suffocated.

心死了，人反而不会死，
像个机器一样活着。

Living without passion turns existence into mere functionality.

不妨大胆一点，反正没人能够活着离开这世界。

Be bold. After all, no one gets out of this world alive.

如果还有很长的路要走,反倒不会太用力。但事实是,我快老了。

Had I more time ahead, I wouldn't strive so intensely, yet time waits for no one, and age catches up.

没必要。
No need for it.

明知会散落，仍不惧盛开。

Let life be beautiful like summer flowers,
and death like autumn leaves.

大彻大悟,步步错,步步悟。

不入歧途,不忘归路。

Profound enlightenment comes from every misstep and realization. Avoiding detours, and never forgetting the way back.

日渐清醒，
得失随意。

**Becoming more aware daily,
letting go of wins and losses.**

只做旁观者,不做局中人。

Be a observer, not a participant.

我的福气在后头,
就是不知道在谁后头。

My good fortune is yet to come,
but I'm not sure where it comes from.

"渔夫在无法捕鱼时,就会修补他的网。"
人生停滞不前的时候,是这句话帮了我。

"When fishermen can't fish, they mend their nets."
This thought saved me during life's standstill.

我学会一边做着力所能及的工作,
一边随心读读书,看看电影,
在这个过程中,
不知不觉就重新有了一张细密的网,
正是这些必要的充电期,
才构成了完整的自己。

I learned to fill my days with small, manageable tasks—
reading books, watching films,
doing what I could.
Slowly, unknowingly,
I wove a finer net for myself.
It is these necessary periods of recharging that make up the complete me.

我是凭着不被他人所知的"那部分"而活着。
人生啊,就是要独自穿过悲喜。

I live guided by the part of me unknown to others.
Life is about walking through joy and sorrow alone.

地球是运动的，
一个人不可能永远处在倒霉的位置上。

The earth is always in motion;
one cannot remain in an unlucky position forever.

一个人只要能满足自己的生活方式就是富有了。
A person who finds fulfillment in their way of life is already rich.

人世间的奢侈品，不过"健康平安"四字。
看淡人间烦恼事，只向心中觅清凉。

The rarest luxury in this world lies in simple words:
health and peace;
Shed the weight of worldly troubles, and seek the inner peace.

我们都是突然长大。
那个瞬间，在无可挽回的事实前，
学会了从容不迫。
在大势所趋时，学会了不动声色。
开始保守地给予，迅速地放弃，
游刃有余地周旋。
在那些众口一词的节日里，
将最好的情感夹杂在寻常祝福中，
试图蒙蔽隐秘的初衷。

We all grow up suddenly.
In that moment, faced with irreversible facts, we learn to remain composed.
When the tide of events is inevitable, we learn to stay calm and collected.
We begin to give cautiously, let go quickly, and maneuver with ease.
During those widely celebrated holidays, we blend our deepest feelings into ordinary blessings, instead of speaking our mind.

自己买单的快乐，才比较保值。

The joy you buy for yourself always feels more enduring.

我永远相信平行宇宙。
可能遗憾的事情都会在
那个时空被弥补。

I have always believed in parallel universes, where perhaps the regrets of this world are mended in another.

健康!

平安

发财.

快乐!

假如生活出卖了我，
我希望是论斤两卖。

If life is going to sell me out,
I hope it does so by weight.

shòu

说出来会被嘲笑的梦想，
才有实现的价值。

A dream that seems unrealistic and even makes others laugh is often the one that holds the value of being realized.

等风来，拨云见日。

Wait for the winds to change; they will part the clouds and let the sun through.

累了就好好休息吧，身体健康最重要。
If you're tired, rest. Nothing matters more than your health.

**稳稳"晴雨表"**

| 开心 | 平和 | 喜欢 | 期待 | 生气 | 惊讶 | 悲伤 | 未知 |
| --- | --- | --- | --- | --- | --- | --- | --- |
|  |  |  |  |  |  |  |  |

正确地开始,微小地长进,然后持续。

Begin with intention, grow in small steps, and persist without end.

要学会经常为自己制造些治愈心情的美好小事。

Learn to create small, healing joys for yourself.

哪怕只是在跨入家门前,去楼下便利店买一支草莓味的冰淇淋。

Even if it's as simple as buying a strawberry ice cream before stepping inside your door.

这是一种非常珍贵的能力,不要轻易丢弃。

This is a rare and precious skill—never let it fade.

曾经，每一次剪短发，都因为有故事发生。

Once, every haircut told a story.

而现在，纯粹是为了节省洗发水，
为了地板上不再有头发丝儿……
为了体重秤上的数字再掉下去二两，
为了让本就琐碎的生活再简单点吧……

Now, I cut my hair to save shampoo, to keep the floor free of stray strands.
To see the scale drop just a little, and to make an already messy life feel a touch simpler.

我们交换故事吧,我有酒。

Let's exchange stories; I've got some wine.

没关系，我会救自己千千万万次。
纵使一地鸡毛，也要重新整装出发！

It's okay. I'll save myself a thousand times if I must.
Even if life leaves me in chaos, I'll pick up the pieces and start again.

在几年前的某一天，我接受了"自己原来只是个平庸的普通人"的事实。
从那一刻起，我的生活轻松了许多，但有种说不清道不明的东西，从此也消失不见了。
嗨，无所谓，快乐就好。

One day, years ago, I accepted the fact that I'm just an ordinary person.
Since then, life has felt lighter—but something indescribable quietly slipped away.
It doesn't matter; as long as you're happy.

快乐秘籍，停止瞎想。
The secret to happiness is stop overthinking.

姑娘你风风火火，究竟是为了什么?

Miss, you're so full of energy; what is it all for?

伸手就能触碰到你的心。

I can reach out and touch your heart.

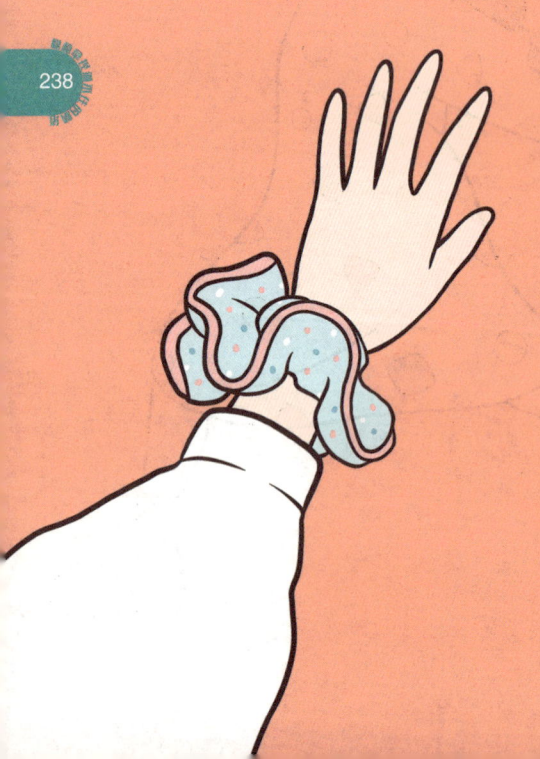

不用理解，各有各的路。

No need to understand—everyone walks their own path.

如果事与愿违,那一定是另有安排。
所有失去的,都会以另一种方式归来。

**If things don't turn out as you'd hoped, it's because something better awaits.**

**What you lose will always return in another way.**

不要在个别人和事上停留太久，别太用力，
降低期待，给生活留点缺口，
那才是阳光照进来的地方。

Don't linger too long on certain people or events.
Loosen your grip, lower your expectations, and leave space in life.
That's where the sunlight finds its way in.

低谷？深渊？
无所谓，下去也是前程万里。

A low point? An abyss?
It doesn't matter—even down there, the horizon stretches on.

此刻小情绪 | 开心 平和 喜欢 期待 生气 惊讶 悲伤 未知

# 熬都熬过来了，我就不回头了。

Having made it through, I won't look back.

不可能天天都是好日子，
有了不顺心的日子，
好日子才会闪闪发亮。

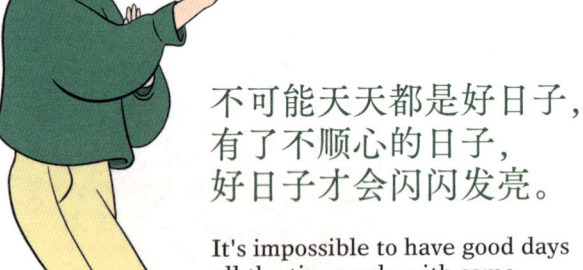

It's impossible to have good days all the time; only with some unhappy days can the good days shine brightly.

心愿花。

**A flower of wishes blooms in the heart.**

勇敢的人先享受世界。人生只是场体验,请你尽兴。

The brave are the first to savor the world. Life is an experience—live it fully.

已走完所有的弯路，接下来都是坦途。

朝朝暮暮，岁岁平安。

I've walked every winding road; all that's left is the open path ahead.

Morning after morning, year after year—may peace follow you always.

从头开始。这一次不是重新开始了，
　　是从经验开始。

Begin again, not from scratch but from experience.

从头开始..

有人接单吗?
接一下我的孤单。

**Can anyone take an order?**
**Please, take away my loneliness.**

"日日是好日" 留言板

过分的忙碌不是勤奋，是奔命。
生命没有往返，过去就是永远。
别因为赶路，忘了爱自己。

Excessive busyness isn't diligence—it's self-destruction.
Life doesn't offer round trips. The past is gone forever.
Don't forget to love yourself while rushing through life.

我把自己贩卖出去，
又再逐渐把自己赎回来，
一生就过完了。

I sell my time to make a living, and with the money I earn,
I reclaim my time to live the life I truly desire.
Just like that, a lifetime passes.

年轻的时候一无所有，唯一有的是
一颗空荡荡的心，和完全不会累的身体。
这两个加在一起，可以一个人造就
一整个世界。

When we are young, we possess nothing but a light heart
and an untiring body. These two qualities empower a person
to create an entire world.

现在，虽然一切变了模样，
也别让结局遮挡了故事的光芒。

Now, though the world has shifted its shape,
let not the closing chapter eclipse the glow of the story.

羡慕一些轻盈的人，
身上不挂只言片语，
只说一些流动的话，
浑身散发着温柔的冷漠。

I envy those who carry themselves lightly—
who let no words weigh them down,
who speak in flowing sentences,
their entire being radiating gentle detachment.

改变不了的事就别太在意，
留不住的人就试着学会放弃，
受了伤的心就尽力自愈，
除了生死，都是小事。

Don't hold on to what cannot be changed.
Learn to let go of those who cannot stay.
Heal your heart, even if it's been scarred.
After all, everything but life and death is just
a small matter.

别人怎么都可以搪塞，自己的心灵却无法蒙混过关。
You can deceive others, but cannot fool your own soul.

想要快乐，我们一定不能太关注别人。
当你看清了自己的样子，
就能吸引到对的人了。

To find happiness, you must stop focusing so much on others.
The moment you see yourself clearly,
you will begin to attract the people meant for you.

你在别人心里是什么样子，
大多时候早就确定好了，
无论做多少努力，最后也只会沦为佐证
他想法的证据罢了。
在乌鸦的世界里，洁白的羽毛都是有罪的。

How others perceive you is mostly already determined.
No matter how hard you try, in the end, you will only
become evidence that proves their assumptions.
In the world of crows, white feathers are always guilty.

你飞得越高，在那些飞不起来的人眼中就越渺小。

The higher you fly, the smaller you appear to those who cannot fly.

# 人从悲哀中落落大方走出来，就是艺术家。

A person who walks gracefully out of sorrow is an artist.

"日日是好日" 留言板

我把委屈讲给雨听，它替我哭了。

I told my grievances to the rain, and it cried for me.

第一次见你的时候，
我的心里已经炸成了烟花，
需要用一生来打扫灰烬。

The first time I saw you,
fireworks exploded in my heart.
It will take me a lifetime to
sweep away the ashes.

世界上最大的勇气，
是压力下的优雅。

The greatest courage in the world
is grace under pressure.

很多年以后,那些"好极了"和
"糟透了"的时刻我们都会忘记,
唯一真实和难忘的是,
我们抬头挺胸走过的人生。

Many years from now, we will forget the moments that felt perfect or unbearable.
The only thing that will stay with us is the memory of standing tall as we walked through this life.

若不是因为你所爱之人居住其中，
这个宇宙没什么大不了的。

**The universe would mean nothing if not for the ones you love who live within it.**

此刻小情绪 | 开心 | 平和 | 喜欢 | 期待 | 生气 | 惊讶 | 悲伤 | 未知

天底下很多事情我们都搞不定，但是，天底下值得做的事情太多了，拣那些我们力所能及的去做，就足够了。

Many things in this world are beyond our control, but there are so many worthwhile things to do. Just pick those that are within our ability, and that will be enough.

当你能做到毫不犹豫地拒绝别人
却不心生内疚的时候,
就是你快意人生的开始。

**The moment you can say no without hesitation or guilt, that's when true freedom begins.**

当你在羡慕别人时，或许你并没留意，
别人也正用羡慕的目光注视着你。

**While you're busy envying someone else,
you may not realize that they're looking at you with the same longing in their eyes.**

如果不是那些倒霉的日子，
好运也不会如此美好可贵。

**Without those unlucky days,
good fortune would never feel so precious.**

别怕美好消失，
我们先让它存在。

Don't be afraid of beauty fading.
Let it exist first, and cherish it while it's here.

我常常发呆，甚至觉得亏欠自己。

I often catch myself lost in thought,
and I can't help but feel like I've let myself dow.

做一个清醒者，冷冷地旁观这世界，
索取自己想要的东西，做自己想要做的事，
成为自己想成为的人。

Be a clear minded observer of the world.
Take what you want, do what you will,
and become the person you wish to be.

将自己的感受高高置顶，不必周旋。

Place your own feelings at the forefront; there's no need for compromise.

向上的生命力，
比皮囊更有杀伤力。

The resilient, upward spirit is far
more powerful than mere beauty.

我已经没有兴趣给每个人留下好印象。
你觉得我是怎样的人,你就配怎样的我。

I've lost interest in leaving a good impression on everyone I meet.
How you perceive me is how I will be to you.

意识流画板

我不在意
你看穿我的细节，
每个人都有瑕疵。

I don't mind if
you see through my
flaws; everyone has
imperfections.

外界的声音只是参考，你不开心，就不要参考。

Outside voices are just suggestions. If they don't make you happy, ignore them.

生活自会消化一切。

Life will digest everything, in its own way.

百味尝遍，自然看透。

**Once you've tasted all the flavors of life, you'll naturally see through its illusions.**

"爱自己"是人生法则，
你的名字必须闪闪发光。

Loving yourself is life's golden rule.
Your name deserves to shine.

279

使人疲惫的不是远方的高山，
而是鞋里的一粒沙。

What exhausts you is not the distant mountains,
but the grain of sand in your shoe.

自在摇曳，生生不息。

Sway freely, and let life bloom endlessly.

此刻小情绪 | 开心 | 平和 | 喜欢 | 期待 | 生气 | 惊讶 | 悲伤 | 未知

允许自己脆弱，
拥有缓慢向上的勇气。

Have the courage to allow yourself
to be vulnerable and to rise slowly.

2:10 AM

失眠？！

落子无悔，不破不立。
无须比较，我就是我。

No regrets with the moves I make.
Without breaking, there can be no rebuilding.
No need to compare—I walk my own path.

你知道吗?
成长就是世界观不断崩塌
又重建的过程。

Did you know? Growing up is the constant breaking and rebuilding of your worldview.

ü 刷新中…

你一定要爱点什么,
恰如草木对光阴的钟情。

You must love something,
just as the grass and trees are devoted to the passage of time.

285

人都是矛盾的，
渴望被理解，
又害怕被看穿。

Humans are contradictory:
longing to be understood,
yet terrified of being seen through.

我可以是任何样子。

I can be anyone I want to be.

休息一下，暂时放过自己。

Take a break, and let yourself off the hook for a while.

"日日是好日"留言板

以大"橘"为重。

Prioritize the big orange—it's what truly matters.

以 大 橘 为 重

会过去的，会到来的，会拥有的。

What's past will pass, what's to come will arrive,
and what's yours will be yours.

你既无青春，也无年迈。
只是像饭后的一场睡眠，把两者梦见。

You do neither youth nor age,
but as it were an after-dinner's sleep, dreaming on both.

允许一切发生。

**Allow everything to unfold as it will.**

没聋，没瞎，有空，懒得理。
我在陪着我自己。

Not deaf, not blind, and very much available—
just choosing to ignore.
I'm spending time with myself.

要学会钝感力，
太敏感的人真的不易快乐。

Learn the art of dullness.
Overly sensitive people rarely find happiness.

虚化所有人群，聚焦你。
Blur the crowds, and focus on yourself.

此刻小情绪 | 开心 | 平和 | 喜欢 | 期待 | 生气 | 惊讶 | 悲伤 | 未知

295

逃跑吧，逃出这虚情假意的世俗。

Run—escape from the mundane world of empty smiles and shallow lies.

有趣有盼，无灾无难。

A life of hope and joy, free of hardship and harm.

一颗自由心脏,
一匹战马坐骑,潇洒出发吧!

**With a free heart and a trusted steed, set out boldly.**

允许停滞和放空，顺从自己，再想其他。

Allow stagnation and emptiness, follow yourself before thinking about anything else.

# 平静，柔和，才气顺。

Calm and gentle, only then will your energy flow smoothly.

## 往前走啊，想什么呢?

**Keep moving forward—why are you hesitating?**

会好，迟早。

**It'll get better. Sooner or later.**

会好　　　迟早

# 但愿新冬，胜旧冬。

**Let the new winter be kinder than the last.**

"日日是好日"留言板

坑填满了，都是路。
人生没有无用的经历。

Once the holes are filled, all that's left is the path.
There are no useless experiences in life.

坑填满了
都是路。

秋裤扎进袜子里，安全感拉满。

**Tucking thermal pants into your socks—
instant security maxed out.**

赶路，也感受路。

As you rush ahead, don't forget to feel the road beneath your feet.

# 你一闭眼世界就黑了，
# 你不是主角谁是?

**The moment you close your eyes, the world goes dark.
If you're not the protagonist, who is?**

稳稳"晴雨表"

| 开心 | 平和 | 喜欢 | 期待 | 生气 | 惊讶 | 悲伤 | 未知 |
|---|---|---|---|---|---|---|---|
|  |  |  |  |  |  |  |  |

不要责怪过去的自己，
她曾经徘徊在雾里也很迷茫。

**Don't blame your past self.
She wandered lost in the fog.**

这一路都是风景，
我就不为你翻山越岭了。

The scenery is everywhere—
I won't climb mountains for you anymore.

最近的状态：
剪不断，理还乱。

My current state:
tangled threads,
impossible to cut or sort.

人生有什么可比的呢！
比美貌我们都会老去，比财富我们都会失去，
比生命我们都会离去，做自己就好。

What's there in life to weigh against another?
Beauty crumbles, etched away by years.
Wealth drifts off like dust on a restless wind.
Life itself—a breath, a flicker—slips beyond our grasp.
All that remains is to live as you are, raw and true.

如果神明还不帮你，
说明她相信你。

If the gods haven't helped you yet,
it's because they believe in you.

且停且忘且随风，
且行且看且从容。

Pause and forget; let the wind carry you.
Move forward with grace, watch with ease, and walk with calm.

比起未来可期，我更喜欢如约而至。

I prefer promises fulfilled over dreams of a hopeful future.

人类的本质是流浪。

**The essence of humanity is to wander.**

是烂片，也要杀青快乐。

Even a bad film deserves a joyful wrap.

"日日是好日"留言板

365个祝福旺自己一整年

为了避免结束,
我避免了一切开始。

To avoid endings, I've avoided all beginnings.

好运常在

"加油"我已经说腻了。
现在我要说:
"祝你拥有随时停留和休息的底气。"

I've grown tired of saying "keep going".
Now, I'd rather say:
"wish you have the courage to pause whenever you need to,
and the strength to rest without guilt."

# 健康平安

时间存在的意义,
就是任何事都不可能立刻实现。

The purpose of time is to remind us
that nothing can be achieved instantly.

# 来自女朋友的排面
## 带你体面"控场"

永远做一个三观比五官正的女孩子。
Be the girl whose values outshine her beauty.

生而独一无二，无须人云亦云。

Born one of a kind—there's no need to follow the crowd.

不追问得不到回复的消息，
不在意不值得的人，
不怀疑当下的真心。

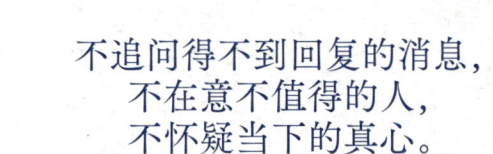

Don't chase after unanswered messages,
don't care about unworthy people,
and never doubt the sincerity of the present moment.

有时候你想证明给一万个人看，
但最后发现只有一个人懂，
那就够了。

Sometimes, you want to prove yourself to ten thousand people,
but in the end, if one person understands you,
that's enough.

从我开始思考接下来
该做什么的那一刻起,
我好像已经从绝望中走出来了。

The moment I started thinking
about what to do next,
I realized I had already stepped out of despair.

你笑的时候，
世界眯成裂缝，
以前我也住在那里。

When you smile, the world narrows into a sliver of light. I used to live there too.

你得学会放手，一开始会有点难受，
之后就会觉得很舒服了，
丢东西是件好事。

You have to learn to let go.
At first, it'll feel uncomfortable, but eventually,
you'll feel relieved. Losing things can be a good thing.

每个人都不一样。
有的是柠檬，有的是苹果，
有的是仙人掌，有的是小树。
我们要做的应该是让柠檬更酸，让苹果更甜，
而不应该让仙人掌长成参天大树。

Everyone is different.
Some are lemons, some are apples,
some are cacti, and some are little trees.
Our task is not to make cacti grow into towering trees but to
make lemons sourer and apples sweeter.

要活得像榴梿一样，不用所有人喜欢，
但喜欢的人很上头。
或者说，香菜？芝士？辣椒？都可以！！

Live like a durian—not everyone has to like you,
but the ones who do will love you passionately.
Or perhaps like cilantro, cheese, or chili peppers—
whatever you prefer!

如果开始讨厌自己,是不是证明该进步了?
我知道痛苦不会自己消失,
它会长久顽固地横在我面前,我必须保持精力,
才能跟难熬的日子对抗到底!

If you start hating yourself, does that mean it's time to grow?
I know pain doesn't fade on its own. It will linger,
stubborn and immovable.
I must conserve my energy to fight through the hardest days.

成长过程中，我最大的自作多情，
就是在做任何事的时候，
都觉得旁人在看着我。
我现在要学会突破这个思维怪圈，
才能真正往前一步。

Growing up, my biggest delusion was thinking people were always watching me.
Now, I need to break free from this thought trap to take the next real step forward.

自知的愚昧，清醒的消沉。

There's wisdom in knowing your ignorance,
and clarity in a quiet surrender to despair.

"日日是好日"留言板

小时候不理解老人晒太阳，一坐就是半天。
长大了才明白，目之所及，皆是回忆；
心之所想，皆是过往；眼之所看，皆是遗憾。

As a child, I never understood why the elderly spent hours sitting in the sun. Now I realize:
what they see are memories, what they think of is the past, and what they watch is regret.

在自己的世界里独善其身。

Protect your peace and thrive in your own world.

焦虑，就是想象力用错了地方。

Anxiety is just imagination gone astray.

人这一辈子，就是活几个瞬间。

A lifetime is nothing more than a handful of fleeting moments.

如果你真的想实现梦想，
最快的捷径便是竭尽所能地过好今天。

If you truly want to achieve your dreams,
the fastest path is to make the most of today.

我要在你平庸无奇的回忆里，
做一个闪闪发光的神经病。

**I want to be the spark of madness that shines brightly in your otherwise ordinary memories.**

有些事，如果无能为力，那就顺其自然。
Some things are beyond your control—so let them be.

可以温柔，可以强势，
凡事没有绝对，看你今天是谁。

You can be gentle or fierce—nothing is absolute.
It all depends on who you choose to be today.

时间有限，
记得要奔走在自己的热爱里。

**Time is short—**
**make sure you're running toward what you love.**

## 这世界太吵，你要把自己照顾好。

The world is too loud—promise me, you'll take care of yourself.

稳稳"晴雨表"

| 开心 | 平和 | 喜欢 | 期待 | 生气 | 惊讶 | 悲伤 | 未知 |
|------|------|------|------|------|------|------|------|
|      |      |      |      |      |      |      |      |

不开心的时候，流泪不如流汗。

When you're unhappy, tears won't help. Sweat will.

你的灵魂是什么形状?

What shape is your soul?

做自己，而不是解释自己。

Be yourself—there's no need to explain who you are.

# 关关难过关关过，长路漫漫亦灿灿。

Every gate may seem hard to pass,
but every one can be crossed.
The long road stretches ahead, shining brightly with promise.

光阴甚短，时来运转。

Time is fleeting, but fortune will turn.

我疯狂收集每一个快乐的瞬间,
用它们回击每一个倒霉的日子。

I collect every moment of joy I can find
and use them to fight back against the bad days.

祝大家拎得清，自由生命遍地开花。

Wishing you clarity, freedom,
and a life where flowers bloom everywhere.

# 所有的为时已晚，其实是恰逢其时。

What feels too late is often right on time.

大胆去追求吧，甭管是什么，
开心就好，大俗大雅。

Chase boldly after whatever brings you joy—
whether simple or refined, happiness is all that matters.

Money

Money

我羡慕三种人，
头发多的，天天拉屎的，还有一秒入睡的。

I envy three kinds of people:
Those with thick hair, those who poop every day, and those who can fall asleep in an instant.

你只是来体验生命的,你什么都带不走,
也什么都留不住。
你能做的,就是不断尝试、收获、感受,
然后放下。

You're here to experience life. You can take nothing with you and leave nothing behind.
All you can do is keep trying, collecting, feeling, and then letting go.

认真长膘的一天。双下巴续费成功！
"小肚，小肚！"
"我在。"

A long and relentless day—Double chin renewed!
"Little belly, little belly!"
"Yes, reporting for duty."

# grow up.

人的一部分能力是会消失的。比如没有人捧场的幽默,吃过很多亏的仗义,不被欣赏的自信,还有得不到回应的爱。这些能力被心酸和新的能力代替。

Some of our abilities fade with time: humor unacknowledged by an audience, generosity learned through loss, confidence unappreciated, love that goes unanswered.

新的能力是保护自己。

These are replaced by new abilities which are to protect yourself.

你发现了吗?
朋友十二画，恋人十二画，
爱人十二画，家人十二画。
好神奇！

Have you noticed?
The Chinese words for "friend," "lover,"
"beloved," and "family" all have twelve strokes.
How magical!

如果没有天赋，那就一直重复。

If you lack talent, just keep repeating.

# 生活，有剥夺，也有馈赠。

Life may strip things away,
but it always leaves you with something in return.

休息了片刻，恍然发现，
原来不努力是这么舒服。
发发呆，抠抠脚，
什么都不用想的状态太滋润啦。

After a brief rest,
I realized how wonderful it feels not to try so hard.
Zoning out, picking at my feet,
doing nothing at all—it's absolute bliss.

我的人生作文，
就算乱写也是满分。

The essay of my life, no matter how chaotic,
is still worth a perfect score.

当你经历过一些事情的时候,
眼前的风景已经和从前不一样了。
你要不顾一切让自己变得靓丽,
即使是在那些糟糕的日子里。

After you've been through certain things, the scenery before you will never look the same.
Even on the hardest days, make yourself shine.

在破茧成蝶之前，做好一只蛹该做的事。

Before you break free and take flight,
do what a grounded soul must do.

喜欢，就不觉得辛苦，
所有人所有事都这样。

When you love something, it doesn't feel like hard work.
This holds true for everyone and everything.

不计划太多,反而能勇敢冒险。

Planning less opens the door to braver adventures.

哎!
现在加的不是班,
是以前逃课时欠的债。

Now I'm paying for skipping classes with overtime.

愿你在阳光下像个孩子，在风雨里像个大人。

**Shine under the sun with the innocence of a child,
and endure the storm with the resilience of an adult.**

是妈妈，但更是女主。

She is a mother, yet she is also the heroine of her own story.

对于人性，多做理解，少做试探。

When it comes to human nature,
choose understanding over testing.

趁我还鲜活，不允许任何人熄灭我。

As long as I'm alive and burning bright,
I won't let anyone put out my fire.

喜欢并不能当饭吃，
但被你喜欢，我会好好吃饭，爱自己。

Love won't feed me,
but your love makes me cherish myself and live better.

这就是学生时代的暗恋吧……
交作业时,作业本叠在一起,
都可以幸福一整天。

This is what a school crush feels like:
Handing in your with your notebook stacked together with the person you like—
and that small detail makes your whole day.

看电影的时候，我心想，
做个大反派多好呀，就不会这么累了。

我为什么这么善良，
每天 800 次共情，
0 个心眼子。
我连伪装成反派的能力都没有。

While watching the movie, I thought, how great it would be to be a villain—life wouldn't be so exhausting.

Why am I so kind, empathizing 800 times a day, with zero cunning?
I don't even have the ability to pretend to be a villain.

无聊的交谈有时近于拷问。
有趣的事物也不用分享给敷衍的人。

Boring conversations feel like interrogations.
Interesting things should never be shared with
people who only pretend to care.

减肥就是，3 天打鱼，362 天晒网。

Dieting is three days of effort, and 362 days of lounging around.

收拾一下杂乱的思绪，
想一想明天要做一个怎样的人。

Gather your scattered thoughts,
and think about what kind of person you want to be tomorrow.

永远有颗少女心，有趣又光鲜。

**Forever keep a young girl's heart—whimsical, yet radiant.**

此刻小情绪

开心　平和　喜欢　期待　生气　惊讶　悲伤　未知

我每天的上班都是为了下班！哈哈！

**Every workday is just a countdown to clocking out—ha!**

"日日是好日"留言板